ISBN-13: 978-1-939722-04-1
ISBN-10: 1939722047

Photos are from the U.S. National Park Service. United States maps are from The National Atlas.

Published by
Technology Management Associates, Inc.
1699 Wall Street, Suite 515
Mount Prospect, Illinois 60016

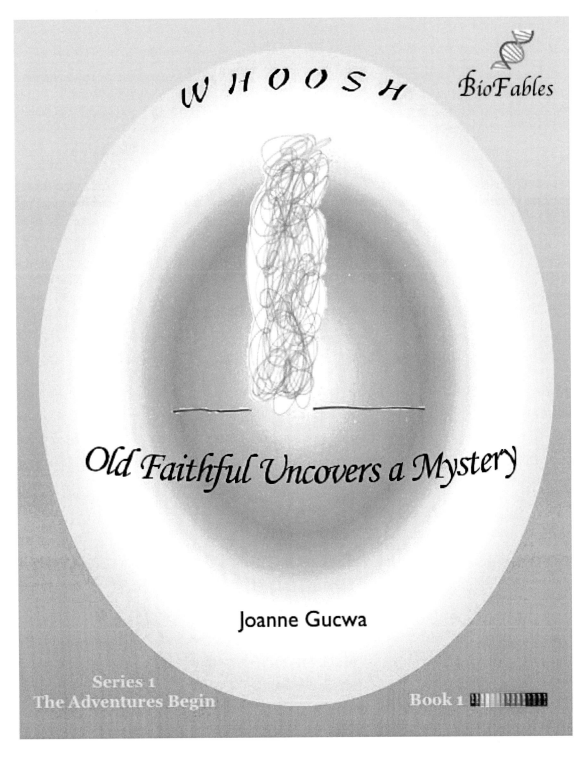

WHOOSH

BioFables

Old Faithful Uncovers a Mystery

Joanne Gucwa

Series 1
The Adventures Begin

Book 1

Published by **Technology Management Associates, Inc.**, Mount Prospect, IL USA

CHAPTERS

Chapter 1
Old Faithful = "Old Faceful?"
Page 1

Chapter 2
Old Faithful Behaves
Page 5

Chapter 3
Bubbling Along the Boardwalk
Page 9

Chapter 4
Nature: A Closer Look
Page 13

Chapter 5
All Steamed Up
Page 19

Chapter 6
How Hot Is Hot?
Page 24

Chapter 7
Living Thermometers
Page 29

Chapter 8
pH? What's That?
Page 35

Chapter 9
Telling Grandpa Mike about Yellowstone
Page 40

Next Adventure
Page 47

HELPFUL HINTS

You are about to join 7-year-old twins, Melody and Mallory Maloney, on the first of many adventures. Melody discovers that she has a special ability that no one else has. Mallory will discover in a later story that he has a different special ability that no one else has. Their special abilities are what make these books *Fables*, that is, "imaginary stories." *Bio* means "life," which can be human, animal, plant and even microbial life. So you can think about *BioFables* as imaginary stories about all kinds of life.

Even though Mallory and Melody have special abilities, they are still very much like ordinary 7-year-olds. They enjoy teasing each other and seem to get into a little bit of trouble once in awhile. Without even trying. Just like you and me.

During their adventures, Melody and Mallory discover a lot of things about nature, science and many other exciting matters. They learn how to look, listen, feel, sniff (but not taste, at least not this time), and think about their new experiences.

Chances are you will find a lot of science and math ideas that are new to you. It's perfectly fine if you decide just to read fairly quickly through the parts that you don't understand right now. You are sure to meet up with these ideas again many times: at home, in school, and in everyday life. The spaces in the outside margin on each page are handy for writing notes and your own ideas.

The Index at the end of this book will help you find the pages where particular words or phrases are used. I encourage you to visit the Index any time you want a fast and easy way to find a section that you would like to re-read.

Go online to visit the BioFables website (www.biofables.com) where you can get more information about this book (and other books in this series). You will find links to other Web sites for extra material that you can use to dig deeper into a topic. This is also the place that you and your parents can find announcements of when new books are available and get other update details.

Each book in this *The Adventures Begin* series shows you a map so you will know where Melody's and Mallory's adventures are taking them. The star shows where they live (a suburb of Chicago near O'Hare Airport) and the circle shows the location of their adventure (in this book it is Yellowstone National Park, which is mostly in Wyoming).

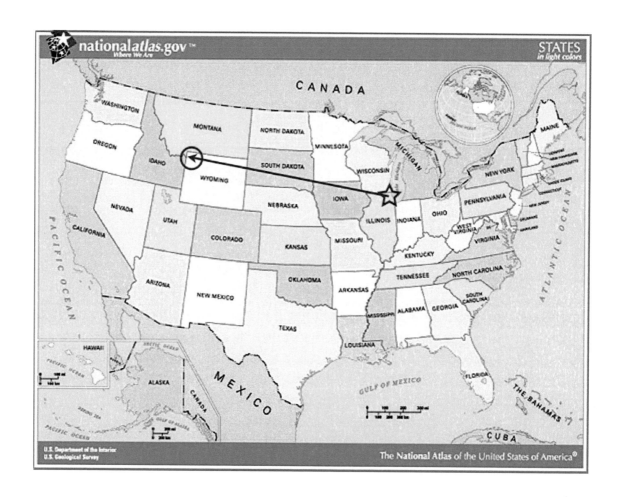

HAPPY READING!

Chapter 1
Old Faithful = "Old Faceful?"

"Race ya!"

The sun was barely peeking over the horizon when Melody shouted this unexpected challenge to her twin brother Mallory.

"No fair. You got a head start!"

A head start was exactly what Melody had in mind. She wanted to make sure Mallory didn't reach Old Faithful before she did. He didn't.

It was far too early for their parents to start exploring Yellowstone National Park. It wasn't too early for the 7-year-old twins, though. They were far too excited to wait for their slow-poke parents.

"Please, can we go out and look around until breakfast? We just want to see if 'Old Faceful' is ready to go off."

"All right, but don't even THINK about going anywhere near that geyser," said their Dad.

"And you do know its name is Old Faithful, don't you?" their Mom, Agnes, reminded them.

"Yay!"

It took only a minute or so for Melody and Mallory Maloney to reach Yellowstone's most famous water fountain. They ran past the wooden viewing benches and almost reached the geyser's cone-shaped mound when the ground began to shake beneath their feet.

Melody thought it felt like a bumpy car ride. Then they heard a gurgling sound. Blurp, blurp. Blurp, blurp. Mallory thought it sounded like a pot of stew simmering on a stove.

The gurgling got louder and louder. Suddenly there was a loud WHOOSH.

"Wow!"

Mallory and Melody had not seen, heard, or felt anything quite like what was happening right in front of them.

Hot steam shot out of the ground and then calmed down. Again and again, the steam shot out, reaching higher and higher.

The steam changed into small water droplets as it came down from so high up. Melody and Mallory felt the cool mist from the geyser as it settled on their faces and clothes.

The geyser was still showing off when a gigantic gust of wind blew a huge wall of steam and super-heated water toward the brother and sister.

Mallory jumped and ran away from the steam's path as soon as he saw what was happening. He remembered his Dad's warning before they left the lodge just a few minutes earlier: "Don't even THINK about going anywhere near that geyser."

Melody just stood where she was. The wall of steam continued to blow toward her. She didn't think at all about running away from the great cloud of steam. It seemed to her as though that cloud of steam knew exactly where she was standing. It was somehow being pushed in her direction.

Melody giggled. "Let's see if "Old Faceful" lives up to the name I gave it," she thought to herself.

Mallory's eyes got wider and wider as the cloud moved toward his sister. Then it completely surrounded her.

"She disappeared!"

Mallory shivered, even though he wasn't cold. He remembered seeing a magician on a TV show who made a lady disappear in a cloud of smoke. It seemed like a long time to Mallory, but it was just a few moments before the cloud of steam evaporated. There was his sister!

Melody's face and hands were covered with tiny droplets of water. They winked and sparkled as the first rays of the sun shined on the droplets. The orange and pink and purple flowers on Melody's jacket looked like they had morning dew on them.

"Whew! Mel's okay," he thought. Just like when the lady in the television show magically reappeared, his sister magically reappeared too!

Mallory's thoughts were completely jumbled. His brain was on high speed, even if his legs weren't. What happened in the steam? His sister wasn't hurt. That's good! Why didn't I run to help her instead of running away to help myself?

Mallory's legs finally got the message and he ran to his sister.

"Mel, are you okay? Why didn't you run away when you saw it coming?"

It took a moment or two before Melody answered. She looked really happy. Some people might say she was 'on cloud 9.' Even after the cloud disappeared.

"That was super, Mal! It was so pretty! A million colors were sparkling inside the cloud!"

OUCH!

Then Melody noticed how worried Mallory looked. In her best teasing voice she asked, "What's the matter with you, Mal? Chicken? Afraid of drowning?"

Melody looked down and noticed little red dots on Mallory's left arm. His arms were bare because he liked the feel of cold. He didn't wear jackets or even long-sleeved shirts unless his parents insisted.

Some droplets of the steamy hot water reached Mallory, even though he was quick to run away. The red dots looked like freckles. Mallory didn't feel anything until now. Now the dots began to sting.

He rubbed his arm, which made it worse. "Ouch! Mel, you're not burned?"

"Of course not, silly," Melody said. "But your arm looks like it's got measles. Does it hurt?"

"It sure pinches a lot. It feels like a billion mosquitos are biting me."

It was Melody's turn to feel relieved. "That's what you get for wearing short sleeves when it's cold outside." She liked to mimic the tone her mother sometimes used when she or her brother would complain about the results of their own poor decisions.

Uh oh. They'd probably get scolded for disobeying their parents' warning about not getting too close to the geyser. Not a good way to start out a family trip.

"We'd better get back so you can change into your green shirt with the long sleeves," Melody said. "Does it still hurt?"

"Nah, it feels better already," Mallory said bravely. He didn't really mean it, though. But he did suddenly feel cold and started shivering again, so they raced back to the lodge.

Just in time! The front door of the lodge opened and there were Mom and Dad.

"What were you two up to?"

"Nothing. Just looking around. Mal's going to change into his other shirt 'cause it's chilly out," Melody answered.

The twins knew how to avoid stern lectures and scoldings from their parents. Besides, they DID stay outside of Old Faithful's boundary so they didn't feel that they did anything so terribly wrong.

It was so confusing! The cloud of steam felt wonderful to Melody, but just a few droplets from that same cloud left spots on Mallory's skin that pinched and burned. How could that be?

Chapter 2
Old Faithful Behaves

Mallory ran quickly to change his shirt. The red spots were already disappearing, but he wanted to keep his arms covered until they completely disappeared.

Their Mom insisted that the family eat a hearty breakfast before starting out. "We've only got two days here, so you need to have good fuel in your bellies to keep your engines running in top form."

"Did you like watching Old Faithful erupt all by yourselves so early in the morning?" their Dad asked at breakfast. "I heard some rumbling and looked out the window just as it started spouting."

"It was neat," Melody said. "It blew all around me and it was all warm and sparkly inside the cloud. Here. Let me draw what I saw inside the cloud."

"Is that so?" their mother said as she looked at Melody's drawing. This was more like a statement than a question. "You have quite a vivid imagination, young lady."

"It's true, Mom," said Mallory. "We were just standing there and all of a sudden, Mel disappeared!"

"It seems you have a vivid imagination too, young man. I was watching from the window in our room and I could see both of you, plain as day," said his Dad.

"But Dad..." Mallory began and then stopped. He didn't want to try to explain what he couldn't understand.

Their parents then asked what they wanted to see most of all. They explored the Yellowstone website at home; now it was time to figure out how to make the best of their time this weekend exploring the real thing.

Melody said she really liked geysers. She also wanted to see and draw pretty hot springs. "Not the mud pots, though. They're ugly!"

Mallory wanted to learn about how the geysers could keep blowing their tops without running out of steam. He borrowed some paper from his sister's notebook and started drawing.

While Mallory was drawing, their mother reminded them that where they were...right now...had more geysers than any place in the whole world. "This is going to be a really special adventure!"

Their Dad agreed. "Besides geysers, Yellowstone also has many unusual living things. Did you know that trillions of tiny microbes live in the hot springs? They give the paint pots their pretty colors. But each one is so little, you'd need a microscope to see them individually."

Mallory and Melody wanted to know more, but they were more interested in starting (or rather, continuing) their adventure instead of just sitting around asking questions.

Old Faithful's Back to Normal

The Maloneys finished breakfast and walked, more leisurely this time, to Old Faithful. It was about time for a repeat performance of one of nature's most magnificent shows.

Mallory and Melody had temporarily forgotten about their mysterious experience. They were just as excited as all the other visitors who had come for the first time to watch Old Faithful's clouds of steam and water shooting high into the air.

Their Mom said it looked like Old Faithful was performing a Water Dance. As she said this, she began moving her arms gracefully up and down. She stood on her tiptoes when she raised her arms and rested back on her heels when she lowered her arms, just like a ballerina (a ballerina in slacks and hiking shoes?). It looked like she was pulling up the powerful forces from far beneath the earth, directing the geyser to perform for the rest of the people who had gathered around. Several children nearby started moving their arms up and down, too. More and more people joined in until nearly everyone was moving in time with Old Faithful's eruptions.

Their Dad smiled and put his arms around both his children. "Your Mom looks like she's leading some of the Tai Chi exercises that she teaches at our Y."

Melody then remembered and poked Mallory. She whispered, "Does your arm still hurt?"

"No, I forgot all about it," Mallory whispered back.

Their Dad asked, "What are you two whispering about?"

"We noticed that Old Faceful is blowing its top just like it did when we saw it before breakfast," Mallory answered.

"It's Old Faithful, you know," their Mom said, without interrupting her dance-like movements.

"We know," Melody and Mallory giggled. "But we like the name *Old Faceful* better."

When Old Faithful's eruptions stopped a few minutes later, everyone around the geyser started clapping their hands. What a fun way to share such a special display with other people!

A Stop at the Visitor Center

There was so much to do and see in such a short time. Even with all the research the Maloney family did before their trip, Yellowstone's choices seemed endless.

The friendly guide at the Visitor Center said, "Lots of people who visit us for the first time come for only a few days, so you're not at all unusual. Where are you folks from?

"We live in a suburb of Chicago, near O'Hare Airport," said the twins' Mom.

The guide said, "Wyoming is far from home. What brings you to the world's largest collection of geysers?"

The twins' Dad explained, "My Dad wants our children to enjoy new experiences, so this is his gift for their seventh birthday. It's our first trip, one each month for a year."

"That's a wonderful birthday present! What are you most interested in seeing?" The guide looked first at the twins.

Mallory wanted to know more about how geysers really work. Melody wanted to visit the hot springs and paint pots. Their mother wanted to find out if Yellowstone's hot springs were used for health treatments, and their father wanted to learn about the bacteria that live where no other kinds of bacteria, or any other kind of plant or animal, could survive.

"What luck," the guide said. "We have a scientist visiting the park this weekend. I'm sure Dr. Shun would be happy to talk with you and answer all your questions. I'll contact her and introduce you. Meanwhile, you can drive over to see the paint pots along the boardwalk. It's not far, and Dr. Shun will probably meet up with you there. Here is a map."

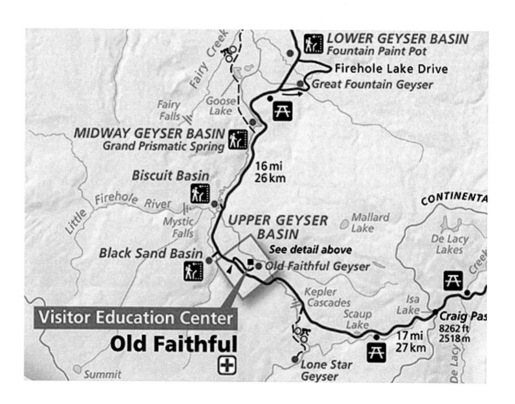

Chapter 3
Bubbling Along the Boardwalk

Old Faithful required a respectful distance, as Mallory had already found out.

The hot springs area was different, though. The nature trail around the colorful paint pots (and not-so-colorful mud pots) allowed visitors to explore the watery pots up close.

The boardwalk looked like it was built in the center of a different planet. It sounded like a different planet, too. It greeted visitors with all sorts of bubbling, burping, hissing, and splashing noises.

And the colors! Melody couldn't decide which she liked best: the deepest blue water you can imagine, the bubbling purple mud, the pinks and yellows and oranges and all sorts of other colors on the ground and on the rocks and in the water. *And ordinary people like us are actually allowed to come into this strange and wonderful place*, she thought.

Mallory, though, felt a little nervous. He wasn't exactly the most graceful boy in his classroom. Walking on the wooden boards of the boardwalk felt odd. It wasn't at all like walking on a sidewalk or in the grass. It felt a little scary to him because there wasn't a street or a house or a lawn on either side of the boardwalk. No handrails, either. Water and steam and rocks were just inches below his feet. He hoped he wouldn't lose his balance and fall in.

Melody loved the feel of the wood. She wanted to reach in to feel the beautiful blue water in one of the hot springs. She wanted to compare it with the friendly feeling of Old Faithful's steamy cloud.

"Don't even think about it!" her father warned when she stooped down to get closer to the water. "The heat and chemicals are strong enough to melt your crayons and pencils. What do you think it would do to your hand?"

Her father's stern voice and her brother's worried look persuaded her not to try this particular experiment.

"Alright, I won't," Melody promised, although she didn't really want to. "Is it okay if I go swimming instead?" Melody didn't like being scolded and always tried hard to make a joke to make everyone forget about the scolding. Her Mom decided to play along.

"That's fne, so long as you wear your bathing suit instead of your flowery jacket. Those flowers would melt, you know," said her Mom.

The Maloneys didn't understand much about this mysterious world of strange colors, sounds, and unfamiliar smells. They wandered along the boardwalk, not quite knowing where to look.

Girl Geysers?

"Mom, what are girl geysers called?" Mallory asked as they were heading back to the parking lot.

"What do you mean by girl geysers? A geyser is a geyser," his mother said.

"Grandpa talked about this movie, Guys and Gals. I thought that maybe girl geysers were called galma'ams."

His Dad laughed. "Son, I think you mean Guys and Dolls. That's an old-time musical. It was very popular in the 1950's and '60's. But what gave you the idea about girl geysers?"

"Well, when you talk to a man, you call him *Sir*, right? And you call a lady *Ma'am*, right? So I thought if there was a *Guy-Sir* there should be a *Gal-Ma'am*."

"What's so funny?" Mallory asked. He felt a little annoyed that his sister and his Mom and Dad all were laughing at him.

"Melody, may I borrow your notebook?" His Mom wrote down the word G-E-Y-S-E-R in Melody's notebook. "You see, *geyser* is spelled with an 'e' and *guy* is spelled with a 'u.' Both 'gey' and 'guy' sound alike but are spelled differently. I like your name for the musical "Guys and Gals" much better than its real name, though, and I think your Dad does, too."

"And no, there is no such thing as a boy or a girl geyser," said his Dad.

Meeting a Scientist

Just then, a lady came up to the family. She had a spiral notebook that was almost the same as Melody's notebook.

"Are you the Maloney family?"

"We are. Could you be Dr. Shun? We heard from a guide at the Visitor Center that you are studying microorganisms here in Yellowstone. I'm Agnes. This is my husband, Mort, and these are our children, Mallory and Melody."

"Hello. Please call me Ethyl," she said as she shook hands with the adults.

"Hello, Dr. Ethyl," Mallory and Melody said at the same time.

Then Mal said, "Please call me Mal, and this is my sister Mel."

"I'm very happy to meet all of you," Dr. Ethyl said. "Melissa at the Visitor Center texted me. She wrote that you had only a few days here and had a lot of questions. I'm always happy to help youngsters -- and their parents too, by the way -- get an idea of the wonders of nature here at the park."

"Thank you very much for your offer, Ethyl. You must be very busy with your research."

"Not at all, Agnes. I appreciate your concern, but this is a weekend, and I should take time off from my work once in awhile. I sometimes get so involved in my research that I forget what day it is. All work and no play can make a scientist pretty dull, indeed! So, let's play 20 Questions!"

"What's 20 Questions?" Mallory asked.

"It doesn't matter, silly," said Melody. "Besides, now that you asked one question, we only have 19 questions left!"

"I'll answer that, Mal," his Dad said with a big smile and a wink. "Then we'll still have 20 questions left. It's a guessing game. Someone thinks of an object and the other players ask questions that can be answered with a YES or NO until they guess what the object is. If no one can guess the right answer after 20 questions, the person who thought up the object wins that round."

Questions and More Questions

"That sounds like fun," Melody said. "Right now I'm thinking about pink and green and blue."

Dr. Ethyl was happy to play along. "I'd guess you're thinking about the paint pots and hot springs, right?"

Melody nodded.

"You've already had a chance to see some of them from the boardwalk, haven't you?"

"Yes, ma'am," said Mal, who was trying to be on his best behavior. "Mel wanted to feel how hot the water was, but Dad wouldn't let her."

"Your Dad was right," Dr. Ethyl said. "Just imagine what would happen if you accidentally fell in. Mel, what did you like best about the paint pots?"

"The pretty colors. It looked like someone spilled paint in the water. That's why they're called paint pots, right? But what causes those colors?"

Dr. Ethyl said, "I'll try to answer all your questions, Mel. Meanwhile, let's talk about what the rest of your family would like to know. How about you, Mal? What did YOU like best?"

"I liked the burping and bubbling best of all," Mallory answered. "What makes that happen?"

"YOU like to make burping noises, Mal," Melody said quickly, before Dr. Ethyl could answer. "What makes THAT happen?"

"Melody, Mallory no more of that," their Mom warned. "We don't want to waste Dr. Ethyl's time with such silliness."

"That's all right. I don't have children of my own, so this is an education for me, too," said Dr. Ethyl.

"Agnes, Melissa's text message said that you're particularly interested in the minerals in Yellowstone's hot springs. And Mort, she said that you have a special interest in the rare bacteria extremophiles that live here. In a way, all your questions are related. I'll do my best to shed some light on some of the things you'd like to know about," Dr. Ethyl said.

Chapter 4
Nature: A Closer Look

Dr. Ethyl noticed Mel's spiral notebook. "I see that you and I use the same kind of notebook. Do you like to draw, Mel?"

"I like to draw and write about things. This is my new notebook for my drawings and things I'm writing about our trip. See my picture of Old Faithful? Grandpa wants to hear all about it when we get back. It's easier to remember if you write things down."

"Very good! That's what most scientists try to do, too. Even when they use cameras and tablets and other electronics, writing and drawing by hand makes it much easier to remember. Sometimes you learn even more because you have to think about what you're writing or drawing," Dr. Ethyl said.

Mallory didn't want Dr. Ethyl to think he didn't like to draw, just because he didn't have a notebook. "I like to draw too, Dr. Ethyl, but I forgot my notebook at the lodge. I borrowed some paper from Mel. Do you want to see my drawing of how a geyser works?"

Dr. Ethyl looked at Mal's drawing. "You had a good time thinking up how geysers work, didn't you?"

Dr. Ethyl then spoke to all of them. "This is a perfect time to take a second, closer look at what you just saw from the boardwalk. When you see so many unusual things gathered in one spot, they tend to blend together in your memory. It helps to go back to look, hear, and smell.

Pools and Hot Springs

Dr. Ethyl led the way back to the boardwalk. She stopped at the first hot spring. It was completely still. No bubbling or gurgling.

"It's such a beautiful blue color. It looks like jewelry fit for a queen," said the twins' mother.

"This one is called Celestine Pool, Agnes. Can you guess why?"

Normally, Melody and Mallory didn't pay much attention to adult conversation. But they were both fascinated by the sights and sounds around them and wanted to learn more, so they listened quietly.

"The word *celestial* has to do with the heavens or sky, doesn't it? I suppose the name Celestine refers to its sky-blue color. It's breathtaking!"

"Yes, you're right, Agnes," Dr. Ethyl said. "Many quiet, still waters reflect the blue sky above. That's why lakes and oceans often look blue."

"Some of these pools have dissolved minerals just like famous hot spring resorts, don't they?"

"That's right, Agnes. Most of Yellowstone's hot springs have a lot of dissolved minerals such as calcium carbonate, silica, and sulfur compounds. Might you be a physical therapist?"

"Some people call me a wellness guide, Ethyl. I teach different kinds of classes and help people understand more about improving their health. I've developed an easy way to remember some steps to good health. I call it 'FEAST,' which stands for Food – Exercise – Attitude – Sleep – Timing," Agnes answered.

"That's wonderful, Agnes. Many people pay little attention to their health, until it's too late. I'd like to learn more about FEAST, especially the last word, Timing. Let's keep in touch."

"Certainly," Agnes said, nodding.

Dr. Ethyl then asked, "Did you know that Yellowstone is the world's first National Park? Since it was established in 1872, its purpose has been to protect its natural wonders for people like us, and for future generations, to explore. No health spas here."

Eye, Ear, Nose Alert!

Dr. Ethyl then walked a short distance along the boardwalk and stopped. She swept her arms around and said "Look, listen, sniff. What do you see? What do you hear? What do you smell?"

The Maloneys began to understand what Dr. Ethyl meant about getting a closer look. The first time they had simply walked and "wowed" at the strange sights and sounds and smells, but that's about all. They didn't really know what was happening all around.

Now, they noticed that every opening in the earth looked different from the others. Some of the openings were watery, others were thicker. Some were steaming, others were bubbling, and a few were spouting water high in the air.

Mallory said the thick ones reminded him of pots of soup. "They don't smell like soup, though. They probably doesn't taste like soup either. And what's that funny smell?"

"That's hydrogen sulfide, isn't it?"

"Yes it is, Mort. Are you a chemist?" Dr. Ethyl asked.

"Yes, I work as a food scientist. Many people think of rotten eggs when they smell hydrogen sulfide."

"What's hydrogen sulfide?"

"Good question, Mal. Let's continue our observations first. Then, we can all go back to the cabin where I stay when I'm doing research here at Yellowstone. I'll make lunch so we can eat and talk about regular water and what makes this water smelly. Then we can draw some pictures about what's happening below the ground here," Dr. Ethyl suggested.

Mallory nodded happily. Food is never very far from his mind. The thought of food also helped him forget his fear of losing his balance on the boardwalk, although he made sure to stay toward the middle of it.

"We're now coming to an extraordinary natural feature at Yellowstone that you've probably never heard of," said Dr. Ethyl.

How Dry is Dry?

They arrived at a noisy, steaming hole surrounded by rocks. Dr. Ethyl asked, "Do you remember this loud hissing sound from when you walked along here the first time?"

"It reminds me that I'll have some ironing to do when we get back home," said the twins' Mom.

"It sounds like when you let go of a balloon by accident when you're blowing into it," said Melody.

"It sounds like the time I forgot to check the water in the car's radiator and it ran dry," said the twins' Dad.

"It sounds like a really mad cat," Mallory said.

Dr. Ethyl smiled. "What great mental images! And all of them have something in common, just like here. What do you think that is?" she asked.

Dr Ethyl's question seemed to be more like an introduction to some interesting information, so everyone waited for her to answer her own question.

"Steam or other gases hiss when they escape out of a hole or outlet. Scientists call these hissing structures fumaroles," she continued. "They have very little water, compared to the others you've seen here."

Dr. Ethyl then explained that fumaroles are like a hot spring that boils off most of its water before it reaches the earth's surface.

"This area is perhaps unique in the whole world because you can see examples of four different kinds of geothermal features right here from this boardwalk. The wettest are geysers. Hot springs have less water than geysers. Mud pots have even less water and fumaroles have the least amount of water."

"Can a hot spring become a paint pot?" Melody asked.

"Yes, indeed, Mel" Dr. Ethyl said. "During the summer, when it's warm and dry, water in some of the pools evaporate, and then the pools turn into mud pots or paint pots."

Aren't springs bouncy?" Mallory asked.

"In this case, the word 'spring' refers to the hot water below the ground that is brought up to the surface. So, in a sense you're right, Mal. You could say that the water 'bounces' up to the surface as it begins to boil," Dr. Ethyl said. "For now, let's look at what's happening *above* the ground. As I promised, later we'll talk about what's happening *below* the ground."

As they walked slowly along the path, Dr. Ethyl pointed out different formations and explained what they were. There were lots of dead trees. Years ago, some of the hot springs had shifted to where the living trees were growing; the heat and dissolved minerals in the hot springs had killed the trees. Then the hot springs had shifted again, away from the trees that couldn't survive those difficult conditions.

Mallory's idea of the game '20 Questions' was to ask about anything he wanted to know. In any case, he figured he had plenty of questions left (and no one was keeping track anyway). He wanted to learn as much as he could from Dr. Ethyl.

"What does geothermal mean, Dr. Ethyl?"

"That's a very good question, Mal. *Geo* comes from the Greek word for earth and *thermal* comes from Greek too. It means heat. So you can think of the word *geothermal* as earth heat."

"Is that anything like a volcano?" Melody asked.

"Yes, it is, Mel," Dr. Ethyl said. "Volcanoes throw out mostly lava, rocks and gases. Geysers spout mostly water. Actually, Yellowstone is sitting on what's called a volcanic caldera. This caldera is sometimes called the Yellowstone Supervolcano because it's so big."

"Supervolcano. That sounds scary," said Melody.

Her Dad said, "You know what's even scarier? Not looking both ways before crossing the street. "

"Huh? Why is that scarier, Dad?" Melody asked.

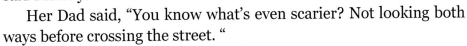

Dr. Ethyl answered, "I think your Dad is trying to tell you that volcano eruptions don't happen very often and scientists track their warning signs so they can alert people who might be in danger. You cross streets all the time, so you have to do your own tracking by looking both ways to know if a car or truck or bus is coming."

That seemed like a very reasonable explanation to Melody. She nodded her head to show she understood.

"Caldera? What's a caldera?" Mallory asked.

"Let me see if I remember," said the twins' Mom, as she recited:

> "*Double, double toil and trouble;*
> *Fire burn, and cauldron bubble.*"

"Macbeth, by Shakespeare," said the twins' Dad.

"Yes, you both have good memories," said Dr. Ethyl. "The word *caldera* comes from the word *cauldron*. A cauldron, like the one from the poem that your Mom recited, is a large, circular cooking pot. It's placed over an open fire instead of on a stove. A *caldera* is a circular indentation in the ground. It's where the earth collapses after a large volcanic eruption."

Hmm, Mallory thought. Volcanos have fire; whatever is in a cauldron (soup?) bubbles over fire. A caldera is circular, a pot is circular.

All of this made perfect sense to Mallory. It also made him hungry!

Until then, no one realized how hungry they were. It was well past lunch time.

When they reached the end of the boardwalk, Dr. Ethyl said, "Please follow me in your car to my cabin. I'll make some sandwiches and we can continue talking over lunch."

Chapter 5
All Steamed Up

The Maloneys soon found themselves at Dr. Ethyl's cabin.

"Thank you for your kind hospitality, Ethyl."

"Not at all, Agnes. I don't get many visitors when I'm here. I'm very happy to have you as my guests...especially since all of you are so interested in science."

Water and Steam

There were two refrigerators in Dr. Ethyl's small cabin. One refrigerator was for food, and the other one Dr. Ethyl used only for the samples she collected from the hot springs. She also had two ovens, one oven for cooking and a separate oven for keeping some of her samples warm. A microscope sat on the middle of her desk.

Dr. Ethyl was happy to hear about the twins' Grandfather's gift of a weekend trip each month. While everyone was munching on apple slices after lunch, Dr. Ethyl said to Mallory, "Mal, you asked about hydrogen sulfide. Let's look at regular water first, since you already know what water is. Hydrogen sulfide looks very similar to water, except for one difference."

Both Mallory and Melody agreed that water sounded like a good place to start.

"Water is a molecule that's made up of hydrogen and oxygen. Have you heard people call water *H-2-O*?" Dr. Ethyl asked.

Melody nodded. "May I please have a glass of H-2-O?"

"That's right, Mel," Dr. Ethyl said. "And water transforms into steam when it's heated and starts to boil. You saw how the steam expands and rises from some of the geysers, compared to the water. That's because the molecules of steam are much farther apart from each other than the water molecules are."

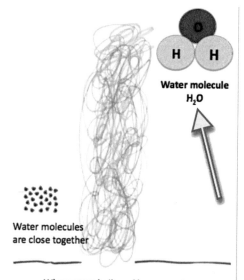

Water molecule
H_2O

Water molecules
are close together

When water boils and becomes steam,
the same number of water molecules
can expand up to 1,500 times.

Melody handed Dr. Ethyl her drawing of Old Faithful. "I'm not sure I know what you mean. Can you draw on my picture, please?" she asked.

Dr. Ethyl drew three circles on Melody's drawing. She explained that water is a molecule made up of *two* parts hydrogen and *one* part oxygen. That's why it is called H_2O.

"You'll notice that I wrote the number 2 slightly below the H and the O," she said.

Dr. Ethyl continued, "Mel, the molecules in the glass of water you just asked for are close together. But when you heat the water and it turns into steam, it expands a lot, and the molecules are much farther apart...up to one thousand five hundred (1,500) times farther apart than they are in a glass of water. On the opposite end, when the water becomes cold and solid, turning to ice, the molecules are much closer together.

1,500 Thumbs

"The difference between hot and cold molecules sounds like a lot. But how much farther apart is one thousand five hundred times?" Mallory asked. He then drew some water molecules in his teapot picture the way he thought they might be as water, and farther apart, as steam.

Dr. Ethyl thought for a moment and then asked Mallory to hold up his thumb.

"Hmm. Just as I thought," she said.

Mallory looked at his thumb. It wasn't dirty. What could Dr. Ethyl mean?

Then she asked, "Mort, how long is the average car nowadays?"

The twins' Dad knew a lot about cars. "Cars are about 15 feet long, on average. Why do you ask?"

Dr. Ethyl took out a small ruler that was in her notebook and measured the width of Mal's thumb.

"About a half inch, I'd say."

Then Dr. Ethyl opened up her tablet computer and found a clip art image of a car. She copied it and pasted it into a drawing program. She made some quick calculations and came up with a neat way to imagine how big 1,500 times something is.

"Think about the width of Mal's thumb as representing how close the molecules of regular water are to each other," Dr. Ethyl said. "Mal's thumb times 1,500 is about the length of four cars. That's how far steam's molecules are apart from each other compared with molecules of water."

 1 thumb = ½ inch

1,500 X ½ inch = 750 inches
750 inches/12 inches/foot = 62 ½ feet

15 feet/car X 4 = 60 feet

"That's neat! How did you do that?" Melody asked.

"Shall I?" asked their Mom.

Dr. Ethyl smiled and said, "Please do, Agnes."

Their Mom explained, "We want to compare a *small* distance to a much *bigger* distance. So, the distance between water molecules is *small* and the distance between steam molecules is *big*, right?"

Both twins nodded.

Their Mom continued: "It helps if you can think of something small that is familiar to you and something else you know that is big. Dr. Ethyl picked your thumb as something small and a car as something much bigger."

Mallory held up his thumb. Melody did too.

"Dr. Ethyl measured your thumb and found that it's about ½ inch wide, right?" their Mom asked. "Next, we want to know how wide 1,500 thumbs are. How do we do that?"

"½ inch x 1,500," said Melody, after thinking about it. "Divide 1,500 by 2. Half of 1,500 is, um, 750, right? Grandpa showed us how to divide the pennies in our piggy bank. Half for me, half for Mal."

"That's right! We also know another number: how long cars are. The number is in feet, not inches, so..." her Mom said, ready to explain further.

"I know, I know!" Mallory interrupted. "The ruler is one foot long and it has 12 inches. So there are 12 inches in one foot. How many feet are there in 750 inches? We have to divide 750 inches by 12 inches in each foot. Dr. Ethyl's picture shows that 750 inches are 62 ½ feet."

"And then we divide 62 ½ feet by the average of 15 feet per car to get approximately 4 cars," said their Mom.

Mallory giggled. "Especially since you can't drive a piece of a car!"

He held up his thumb again. "If water molecules are as far apart as my thumb, then steam molecules are so far apart that you could park four cars in between each one, right?"

Dr. Ethyl was very pleased. "That's exactly right, Mal. Water molecules are really small, though, so they're much closer together than the width of your thumb. In fact, they're so small that you can't even see them through a regular microscope."

Mal nodded. After all that arithmetic, he wasn't going to ask about how small the steam molecules are. He just hoped he'd remember the neat trick for comparing big things and little things.

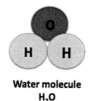

Water molecule
H_2O

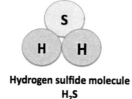

Hydrogen sulfide molecule
H_2S

Dr. Ethyl said, "Mal, you asked about hydrogen sulfide at the boardwalk. Let's compare the molecules of water and hydrogen sulfide. They look almost the same, don't they? Except that the oxygen in water (red) is replaced by sulfur (yellow) in hydrogen sulfide."

"So sulfur is what makes the water smell stinky?" Melody asked.

"That's right. There are other gases in the water, too," Dr. Ethyl explained, "but hydrogen sulfide definitely is the most smelly."

Then Dr. Ethyl had an idea.

"It's much too nice a day to stay indoors. Let's take our notebooks and examine the bubblers and steamers that are all around my cabin."

That sounded like a good idea to everyone.

"By the way," Dr. Ethyl said, interrupting her thoughts for a moment. "This area is secret. The park service makes sure it's hidden. That's why it is not marked on any tourist maps. I need to ask you not to let anyone know where it is. But you know where it is because you are my special guests."

"This is quite a privilege to be your guests, Ethyl, and we promise to keep this place a secret. We really appreciate your taking all this time with us," said the twins' Dad.

"Not at all," Dr. Ethyl said. "There are a few more exciting things I'd like to show and tell you about. AND, we still haven't talked about how geysers work, so let's do that, too. Mel and Mal, are your brains about to explode with all this information?"

Chapter 6
How Hot Is Hot?

Dr. Ethyl led the Maloney family to an area near her cabin where there were several small hot springs and mud pots.

"Be careful to step only on the big flat stones," she warned.

"Uh, oh," Mallory thought to himself. He moved very cautiously, taking great care to step as close to the middle of each stone as he could. No one else seemed to be having so much trouble, though. "You can do it," he told himself as he shakily put one foot in front of the other.

Some of the mud pots were thin and watery, like chocolate milk; others were almost as thick as peanut butter. A few bubbled quietly, as though they were talking to themselves. Others seemed like they wanted to be noticed as they splashed up gops of mud.

As everyone stepped carefully from stone to stone, Dr. Ethyl explained once more that the bubbling is caused when there isn't enough space for the hot water in the ground.

"When the pressure builds up high enough, the water turns to steam. The molecules move faster and take up more space. When there's nowhere else to go, the steam and water shoot out of the ground. That's when it's called a geyser."

"That's like our pressure cooker," Melody said.

"Yes, it's very similar," said Dr. Ethyl. "When water becomes superheated, it escapes as steam from a pressure cooker, or it shoots out of the ground as a geyser. Superheated water can be almost twice as hot as boiling water; it's around 400 degrees Fahrenheit before the air cools it down."

Dr. Ethyl made a quick drawing, showing how much hotter superheated water can be, compared to boiling water.

"No wonder I got burned," Mallory thought to himself. "But I still wonder why Mel was okay."

Dr. Ethyl pointed to her drawing. "Scientists use the Celsius scale. That's on the right side of the thermometer with the letter C. You'll be learning about different temperature scales in your science classes."

"So, if boiling water can burn you, think about how much more dangerous superheated water can be, whether it comes from a pressure cooker or a geyser," Dr. Ethyl advised.

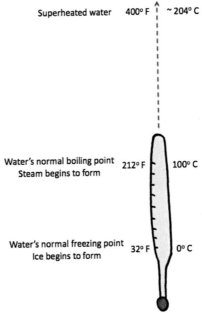

Melody and Mallory nodded. They both wondered if Dr. Ethyl knew about what happened at Old Faithful. It seemed like a long time ago, even though it was just this morning.

"By the way, Mal..." Dr. Ethyl began.

"Uh oh. She knows," Mallory and Melody both thought to themselves.

Instead, Dr. Ethyl said to Mallory, "Your teapot picture is a good start to understanding how geysers work."

"It is?" Mallory asked, surprised.

"It is?" his sister Mel asked, even more surprised.

Geyser Ingredients

"Let's go see one of my favorite mini-geysers. It's right in back of the cabin. I call it Sir Splashalot."

"That's a funny name. Why do you call it Sir Splashalot?" Melody asked.

"Because it sits at a round table in the ground," Dr. Ethyl answered, with a smile and a wink to her parents.

"Sir Lancelot was a knight of King Arthur's Round Table," her Mom explained. "Let's read the book together when we get home."

Dr. Ethyl led them to her cabin's back porch. A few feet away, Sir Splashalot was merrily blowing a small fountain of water into the air. The ground around Sir Splashalot looked like a round table, just as Dr. Ethyl described.

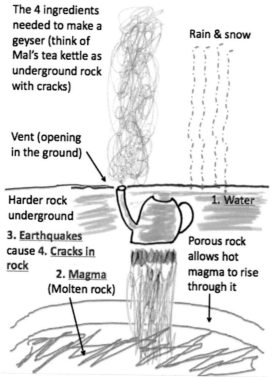

The 4 ingredients needed to make a geyser (think of Mal's tea kettle as underground rock with cracks)

Rain & snow

Vent (opening in the ground)

Harder rock underground

1. Water

3. Earthquakes cause 4. Cracks in rock

2. Magma (Molten rock)

Porous rock allows hot magma to rise through it

Dr. Ethyl drew a combination of two pictures, Melody's Old Faithful on top and Mallory's 'teapot geyser' beneath it. Then she added these ingredients, explaining as she drew:

First, *above* the earth, you need enough rain or snow to fall and seep through the ground so that there is plenty of water *under* the earth.

Next, you need magma. Magma is liquid rock. Hot gases deep inside the earth melt any rock that happens to be present. Can you imagine how hot those gases are that they can melt something as hard as a rock? The magma cools as it seeps up through porous rock until it comes to a harder rock layer that is closer to the earth's surface.

Third, you need earthquakes and cycles of heating and cooling. These cycles cause the rocks to crack. They shift around and gaps form as they shrink when they cool. All of these movements create:

Fourth, a *plumbing system* made up of cracks, called vents, in the harder rock. These cracks are small enough so that the superheated water can't escape so easily.

You remember that steam molecules are much farther apart than water molecules. Those farther-apart steam molecules are what cause pressure to build up. The steam works its way to the earth's surface, finds a vent, and explodes in an eruption of steam and water.

"So, if you take away the rocks, you have a hot spring?" Mallory asked.

"Excellent thinking, Mal!" said Dr. Ethyl.

Melody wasn't about to have her brother get all the praise. "And if you take away some of the water, you get a mud pot or, if the water dries up, you get that, um, that fuming thing."

"Fumarole, isn't it?" their Dad asked.

"That's right, Mort," said Dr. Ethyl. Then, to Melody and Mallory, "You're both starting to think like real scientists!"

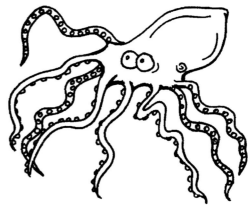

"Agnes, Mort, would all of you like to join me at Octopus Spring early tomorrow afternoon?" Dr. Ethyl asked. "I've been collecting research samples at Octopus over many years. It's a perfect place to see extremophiles. Extremophiles are those colorful bacteria that are able to live in these hot springs."

"Can we, Mom?"

"Can we, Dad?"

Mallory and Melody had visions of a big octopus swimming in a rainbow-colored hot spring. In fact, though, Octopus Spring got its name because water flowing out of the main pool looks like the arms of an octopus.

"Thank you, Ethyl. That sounds wonderful," said the twins' Mom.

"This way you'll also have time to explore on your own. There are some magnificent sights nearby. One you shouldn't miss is Steamboat. It's supposed to be the world's tallest active geyser."

"The world's tallest active geyser," the twins' Dad repeated as they headed out to their car. "Hmm."

"Hmm what, Dad?" Melody asked.

"I feel a poem coming on," he said. "Good thing I brought my notebook too. I'll write it out, play with the words a little to make it a real poem, and when it's ready, I'll treat you to a dramatic reading. Or maybe even a song."

After dinner, the twins' Dad brought out his notebook. "It's just a short poem about geysers," he told them. "You could sing it to the tune of "Twinkle, Twinkle Little Star." And he sang:

> Water erupts from beneath the rock,
> Timing obeys its internal clock.
> It shoots, it sprays, high into the sky.
> Tick and tock, and then the rock's dry.
> Over and over the cycle goes 'round
> And it always makes that whooshing sound.

Mallory said, "That's neat, Dad. Good poem. Fits the song."

Melody thought for a moment. She liked listening to her Dad sing. People were always saying that he had a fine Irish tenor voice.

"Good for cooling soup," he always replied.

His song sounded familiar to Melody. "Isn't that the same tune as *Baa Baa Black Sheep*?" she asked.

Chapter 7
Living Thermometers

The narrow trail to Octopus Spring opened up to an explosion of brilliant colors. Swirly patches of yellow, pink, green, orange, brown and even black decorated the ground around the steaming, deep blue water.

"What incredible colors! Makes me want to repaint all our walls when we get home."

"You do that, Mort. When you do, I'll replace all our windows with stained glass," said the twins' Mom.

"Wow, neat!" Mallory said. But Melody just shook her head. She was pretty sure that their parents were just teasing and having fun.

"Yum," Mallory said, when he realized that his parents weren't serious. The bright colors resembled his favorite fruits and vegetables: blueberries, strawberries, yellow and green peppers.

"Don't even THINK about tasting anything," his Mom warned.

"This is one of the most spectacular shows put on by extremophile bacteria, and they decorate Octopus Spring," said Dr. Ethyl. "Do you see why it's called Octopus Spring?" she asked.

Mallory pointed to the steaming blue pool of water. "The head of the octopus is the round part, right?" Then he pointed to some of the little streams of water flowing out from the center of the pool. "And those are its tentacles."

Dr. Ethyl clapped her hands happily. "That's exactly right, Mal. The water is almost boiling in the middle, at the head of the octopus. As the water overflows from its head into its tentacles, it cools off until it's about the temperature of a cup of hot cocoa. Can anyone guess how you can tell the temperature of the water here without a thermometer?"

At first Melody thought Dr. Ethyl was asking a trick question. She though a moment and then she had an idea.

"The different colors?"

Dr. Ethyl clapped her hands again. "Yes, applause to you too, Mel. You and Mal are both learning how to look at nature and then think about it in a new way. These extremophile bacteria that grow in hot conditions also happen to have different colors!"

Melody felt good that she had guessed correctly. "I like the pink ones," she said. "They're so pretty. Pink is my favorite color."

"Scientists call the pink microbes *Aquifex aeolicus*, but you don't need to remember their name. The important thing to remember is that they can live in water that is very close to boiling. So, wherever you see the pink colors of these extremophiles at Octopus Spring, these living thermometers will let you know that the water is very, very hot," said Dr. Ethyl.

What's an Extremophile?

"But I still don't know what an extreme ... um ... extremophile is," Melody said.

"Let's take the word apart," Melody's Mom suggested.

"It sounds like your Mom is about to demonstrate why we call her the *walking dictionary*," their Dad said with a smile and a nod to his wife. "Please do, Agnes."

"Extremophile. It comes from both Latin and Greek, which is quite unusual. Many words we use have either Latin or Greek roots, but not both. You know what *extreme* means, right? It comes from Latin and it means 'conditions that are far out of the ordinary'. The *phile* part comes from Greek and means 'love'. You've heard of Philadelphia. The city was named *Brotherly Love* by its founders."

"I LOVE Philadelphia cream cheese. EXTREMELY!" said Mallory. Everyone laughed.

Dr. Ethyl then opened her notebook to the thermometer she drew earlier. She flipped it sideways and started adding to the drawing. "This might help you understand what your Mom explained about where the word 'extremophile' came from," she said.

"Most bacteria like temperatures that are in the red square section of the thermometer, between a comfortable room temperature and when you have a fever. That's just a small part of our entire thermometer. It measures temperatures from freezing to boiling."

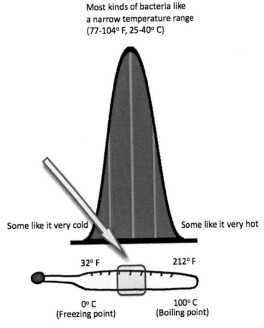

Most kinds of bacteria like a narrow temperature range (77-104° F, 25-40° C)

Some like it very cold

Some like it very hot

32° F

212° F

0° C
(Freezing point)

100° C
(Boiling point)

Mal and Mel liked the ruler-type thermometer better than the digital one they had at home. Different temperatures are much easier to understand when you can see how far apart they are.

Dr. Ethyl continued, "A very few types of the world's bacteria live at the extremes of heat and cold. That's why they're called *extremophiles*. And the extremes are not only hot or cold. Some bacteria live in very dry, very salty, or other extreme conditions. A distribution curve like this dark blue one helps to see what's a normal temperature range for most of the world's bacteria. Oh, and by the way, the extremophile bacteria that like extreme hot conditions are called *thermophiles*. But don't worry about that. The word 'extremophile' is perfectly suitable."

Melody and Mallory weren't entirely sure they understood everything Dr. Ethyl said, but they nodded their heads to let her know they were interested in learning more.

Then Mallory began thinking about the big cloud of hot steam that blew all around his sister. Could his sister be an extremophile? Why wasn't he an extremophile too? Just a couple of droplets made red dots on his arm and felt like needles sticking him. Why did the steam feel so hot to him but not to his sister?

"Bacteria? They're really little, right? Dad told Mel there are tons of them and that's what makes the colors."

"That's right, Mal," Dr. Ethyl said. "There are millions, billions, and even trillions of extremophile bacteria that create this rainbow of colors. They aren't as small as water molecules, though. You can see them through a microscope. Would you like to see them? Here, let me show you."

Under the Microscope

Dr. Ethyl took out some things from the big waterproof bag she brought with her. She set up a little tray with folding legs on the ground. Then she took out a small microscope and placed it on the tray.

"This is called a field microscope," she said. "It's not as powerful as the one on my desk at the cabin, but it's good enough for my outdoor research."

Then she took out a box of glass laboratory slides, an eye dropper, and some colored liquid.

"Some microorganisms look nearly transparent under the microscope; this colored stain helps us to see their internal structure more clearly. The bacteria colors should show up fine, though, so I don't think we'll need the stain today."

Melody pointed to the eye dropper. "Are you going to pick up some extremophiles with that?"

"Yes. Which color would you like me to show you?"

"I like the pink and yellow ones."

Mallory said, "I think the green and black ones are really neat."

"The yellow ones are closest to us so let's pick up some of those," Dr. Ethyl suggested.

She went to the edge of the spring, squeezed the bulb of the eye dropper and slowly released it. Yellow bacteria, mixed with hot water, got suctioned up into the eye dropper.

"Now let's see what they look like under the microscope."

Dr. Ethyl placed one drop of the hot yellow liquid onto a glass slide and placed another glass slide on top "so they don't escape," she said with a smile and a wink.

Melody and Mallory took turns looking into the microscope. This picture is something like what they saw:

"But how can they live? Is their skin so thick that the heat or chemicals can't hurt them?" Mal asked.

That's a really good question, Mal," said Dr. Ethyl. "We scientists still have a lot to learn about extremophiles. But I can tell you for sure that they don't have skin that's like ours."

"Why do you study extremophiles, then?" Melody asked.

"There are a lot of reasons to study nature, Mel. As a scientist, I get excited about discovering something new, something that no other person on earth has ever found out before. And sometimes these discoveries show us ways to heal people who are sick."

"What did you find out about the pink ones?" Melody asked.

"What's really interesting about the pink ones is that they eat hydrogen and need only a very small amount of oxygen to survive. Some companies are using them to improve how paper is made and some are used in foods. I'm sure your Dad will be happy to tell you about how bacteria are used in food," Dr. Ethyl said.

"How about those green ones?" Mallory asked, pointing to a bright streak of green.

"They are a type of cyanobacteria – *cyan* from the Greek word meaning greenish-blue. They live in conditions that are only slightly cooler than what the pink ones live and grow in," Dr. Ethyl said.

"It's still really hot," Mallory pointed out.

"That's right, Mal. They have very special powers, though. During the daytime, they act like plants. They use light to produce energy, just as the leaves of plants do. But at night, they flip some internal switches on and off to start converting the nitrogen gas in the air into proteins," Dr. Ethyl explained.

Then Dr. Ethyl interrupted her own thoughts. "But this is all quite complex. This is the kind of science you'll learn in high school and college."

As Dr. Ethyl was talking, an idea suddenly came to Melody. It was all about the pink and green colors.

"If colors of different bacteria can show how hot the water is, what if we painted a thermometer to match their colors so we'd know how hot the water is. Pink would be on top, then green, then...what color comes next, Dr. Ethyl?"

"Mel, your idea of a color temperature guide would be perfect right here at Octopus Spring. Remember, extremophiles are extremely rare, so you couldn't use it at home or even in different parts of Yellowstone," said Dr. Ethyl.

"Oh." Melody now felt *extremely* disappointed.

Dr. Ethyl reassured her. "A color chart is an excellent idea, though. As a matter of fact, scientists do use different kinds of color charts. I happen to have one in my bag."

Chapter 8
pH? What's That?

Dr. Ethyl took out a small plastic container. It had strips of paper inside and a color chart on its cover. She took out a strip of paper.

The chart is pretty, but the paper looks pretty ordinary, Melody thought.

"This is pH paper," Dr. Ethyl said. You'll learn about pH in high school. For now, though, you can think of it as a scale, just like a ruler and the kind of thermometer I drew earlier."

Dr. Ethyl explained that the scale goes from 0-14.

"So 0 and 14 are its EXTREMES, right?" Mallory asked.

"That's right, Mal. Pure water and milk are in the middle of the scale, at pH 7. The acid in your stomach is on the extreme lower, or ACID, side of the scale. It's got a pH of about 1. On the other side of the scale are many soaps and cleaning liquids such as bleach, which has a pH of about 13. That side is called BASIC, or alkaline.

"If you drop some liquid on the paper or dip the paper into a liquid, the paper changes color according to its pH. You can then compare the paper's new color with the colors on the chart. The chart will show the pH of the liquid. Here. Let me show you how pH paper works."

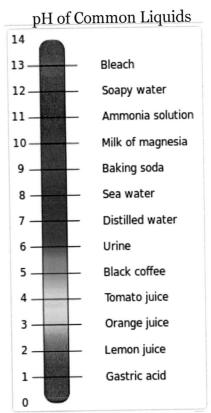

pH of Common Liquids

pH	Liquid
14	
13	Bleach
12	Soapy water
11	Ammonia solution
10	Milk of magnesia
9	Baking soda
8	Sea water
7	Distilled water
6	Urine
5	Black coffee
4	Tomato juice
3	Orange juice
2	Lemon juice
1	Gastric acid
0	

Dr. Ethyl dipped the strip of pH paper into a little stream of water flowing out of Octopus Spring. The paper turned a kind of purple.

"Let's compare the paper's color with the chart," Dr. Ethyl said.

"It's 8!" Melody and Mallory said at the same time. The paper's change in color seemed like magic.

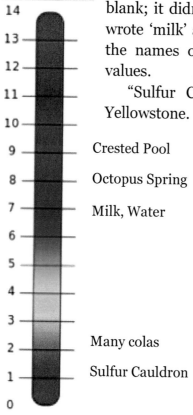

14	
13	
12	
11	
10	
9	Crested Pool
8	Octopus Spring
7	Milk, Water
6	
5	
4	
3	
2	Many colas
1	Sulfur Cauldron
0	

Dr. Ethyl found another color chart in her bag. This one was blank; it didn't have a list of liquids, like the first chart had. She wrote 'milk' and 'water' next to pH 7. Dr. Ethyl also wrote down the names of some of Yellowstone's springs next to their pH values.

"Sulfur Cauldron is the most acidic hot springs here in Yellowstone. It's pH is just over 1, near the pH of your stomach juices. Crested Pool is about the most alkaline, although many chemicals used for cleaning are even farther up on the scale." Then Dr. Ethyl wrote where bottled and canned sodas would appear on the chart.

Melody noticed that colas were really far down on the scale. Mallory also noticed. Melody pointed at the orange color and looked at her mother.

"Now you know one of the reasons why your Dad and I want you and Mallory to drink your milk after meals, and to drink water other times when you're thirsty," their Mom said.

Dr. Ethyl gave the pH chart to Melody along with a bunch of paper strips so she and her brother could test and compare different liquids once they got back home.

Melody then noticed something unusual in the writing on Dr. Ethyl's first chart.

"How come there's a little 'p' and a big 'H' instead of the other way around?" she asked. "Sentences usually start with a capital letter, don't they?"

Dr. Ethyl remembered that the twins' Dad used chemistry a lot in his work. "Mort, you can probably explain better than I how pH came to be written the way it is. Would you please?"

"I'll try, Ethyl," he said and pointed to the pH label. "Mel, most people think of pH as 'power of Hydrogen.' Hydrogen is an element, in fact the first element in the Periodic Chart. The rule is to capitalize the first letter of an element's name. But that's a whole lot of chemistry that we should leave to another day, right Agnes?"

"Yes, indeed, Mort," said their Mom as she looked at her watch. "This day has been filled with new ideas, and we don't want to forget what we've already learned. It's probably a good time for a quick review before we need to head back to the airport."

"I agree," said Dr. Ethyl. She turned to Melody and Mallory. "You've seen a few of the useful things that the different colors of Yellowstone's waters can show us. How many do you remember?"

"Temperature!" Mallory quickly said. He couldn't forget what Melody told him about all the colors she saw when Old Faithful's steam surrounded her.

"Extremophiles!" Melody said at almost the same time.

"pH?" Mallory said next, almost like a question because he wasn't entirely sure what pH really meant.

"Excellent!" said their Dad. "And besides extremophiles, what else causes the colors to be different in different springs? Do you remember what Dr. Ethyl told Mom about the blue color when we were walking on the boardwalk?"

Melody and Mallory shook their heads "no."

Melody looked at the pH chart she put in her notebook and saw the words *Sulfur Cauldron*. "I know, I know! Chemicals!"

Colors can be kinda complicated, can't they?" Mallory asked.

Dr. Ethyl agreed. "That's true, Mal. As we've seen, the water out here in nature comes in many colors. You remembered about temperature and pH and your sister remembered extremophile bacteria and chemicals. All of these things influence the water's color. And remember that the color of the sky is also reflected in the water. Understanding colors in nature certainly can get complicated."

"Complicated is okay," said Melody. "That's what makes everything so interesting."

Time's up Already?

It was nearly time to drive back to the airport.

"Well, Mel, Mal. Do you have any ideas about what you want to be when you grow up?" Dr. Ethyl asked.

Mallory knew exactly what he wanted to be.

"I'm going to be an explorer. I want to go places and be the first to find out things that no one else knows about."

"How suitable that you want to be an explorer," Dr. Ethyl said. "Did you know that a man, whose *last* name was Mallory, was a famous mountain climber? It hasn't been proven, but many people believe George Mallory was the first person to climb to the top of Mount Everest. Mount Everest is the world's highest mountain."

No, Mallory Maloney didn't know about George Mallory. But he was really excited by what Dr. Ethyl said. The first person to climb the world's highest mountain! He decided that from then on he would introduce himself as Mallory, not just Mal. His sister could still call him Mal, of course, and his Mom and Dad and Grandpa Mike could, too, but now he wanted everyone to know that his *real* name was Mallory. "Please call me Mallory, Dr. Ethyl," he said.

Mallory thought about asking Dr. Ethyl how high Mount Everest was. He wanted to try and figure out how many cars it would take, standing on end, to reach the top of Mount Everest.

Melody interrupted her brother's thoughts. "Being an explorer sounds like a lot of fun, but I'd really like to find out more about interesting stuff like those pink extremophiles," she said. She remembered all those pretty colors reflected in Old Faithful's steam. "I also like music a lot, so please call me Melody." She thought her brother's new respect for his full name was a good idea for her name, too.

"You both have wonderful dreams for the future, Melody and Mallory," said Dr. Ethyl. "Keep studying and learning. We have a world full of questions that need to be answered. I have the feeling that both of you will find answers to some of those questions. I also suspect that this won't be your last encounter with extremophiles."

Melody agreed, nodding her head. "Uh huh. That's because *I'm* an extremophile!"

Her Mom, Dad, and Dr. Ethyl laughed, but Mallory didn't think it was funny at all. His sister was an extremophile, and he wanted to be an extremophile, too!

"I don't think so, dear. Only bacteria are extremophiles," Dr. Ethyl said.

Both Melody and Mallory were sure that people could be extremophiles, but they didn't want to argue with Dr. Ethyl.

What a great first adventure! Yellowstone National Park (at least the part the Maloney family saw) was as extreme as the extremophiles that live there. The extreme number of discoveries and things to think about could have filled up a whole month, rather than just two days.

But these two days had been filled up enough, and the whole family slept on the plane most of the way back home.

Mallory and Melody were eager to tell Grandpa Mike all about their trip when they returned home. They organized their drawings, maps, and other items they collected to show Grandpa Mike and to help them remember this exciting trip.

Chapter 9
Telling Grandpa Mike about Yellowstone

Grandpa Mike came over one evening a few days after the family returned home from Yellowstone.

"So tell me, what did you like best?" Grandpa Mike asked, even before he got to his favorite chair in the living room.

"Old Faceful," both twins said at once.

"Old Faceful, eh?" Grandpa Mike smiled as he sat down. "Did you get close enough to get a faceful?"

"A big wind whooshed a whole bunch of steam into Melody," Mallory blurted out. Oops, he thought. He realized that Grandpa Mike might not believe him either.

"But she's okay," Mallory quickly added.

"Oh? What did your parents do? Were they worried?"

"No, Mel and I got up early and went out before them. They were looking out the window from our room, but they didn't see what happened."

Melody then said, "It felt warm and snuggly, not hot at all. But it burned my little chicken-y brother. Just a little, though. He didn't even know when the red spots went away."

"I'm happy to hear that, Melody. What else can you tell me about Yellowstone," Grandpa Mike asked.

Melody and Mallory didn't know it at the time, but Grandpa Mike felt that this strange happening with Melody might be the first sign that his grandchildren may have special abilities that other people didn't have. He wasn't sure how he knew, but he knew.

Some Chemistry

"We learned what causes lakes and oceans to look blue," Melody said. "Here's one of Mom's pictures. It's a hot spring."

Mallory and Melody hadn't taken any pictures. "Cameras sometimes have a way of getting in the way and distracting from real life. We want you to fully enjoy your new experiences," their parents said. Their Mom and Dad did take a few photos for sharing with others and to help them remember their visit.

"Why do the hot springs look so blue, Melody?" Grandpa Mike asked.

"Sometimes it's because of the blue sky, and sometimes it's from bacteria and sometimes it's from chemicals in the water."

"And sometimes the chemicals make them smelly," said Mallory, "like…"

Mallory quickly found the page in his own notebook and said proudly, "hydrogen sulfide! We even learned that hydrogen sulfide is H_2S and water is H_2O."

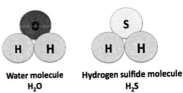

Water molecule H_2O

Hydrogen sulfide molecule H_2S

"I'm very impressed with your knowledge of chemistry," said Grandpa Mike. "How did you learn all this?"

Melody and Mallory told Grandpa Mike that they met this really nice scientist, Dr. Ethyl, who explained lots of things to them. Not that they understood everything that Dr. Ethyl said.

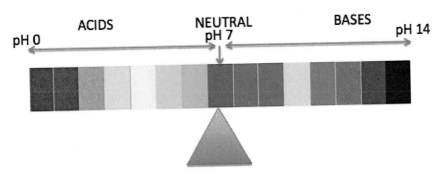

"Dr. Ethyl taped this pH chart in my notebook. It was up and down, but I like it better sideways," Melody said. "Then it's like a teeter-totter. pH 7 is in the middle and the teeter-totter doesn't tip. That means water and milk are balanced. Not acid, not base. Dad drew the lines and added the words and numbers."

"We learned lots more chemistry, too," Mallory said as he looked through his notebook. "Grandpa, did you know that steam is just water that's really hot? Sometimes steam is even hotter than boiling water."

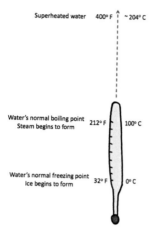

Before Grandpa Mike could answer, Melody opened her notebook to the thermometer drawing that showed superheated water. She pointed to the thermometer. "Scientists use different thermometers than we do. Ours are Fahrenheit, theirs are called Celsius, or sometimes Centigrade. I think Celsius is easier than Fahrenheit. Freezing is zero, boiling is 100. See that squiggly line next to the hottest number? It means that the number is not exact, but it's close enough."

Neat Ways of Thinking

Mallory found his teapot drawing showing water molecules close together and steam molecules farther apart.

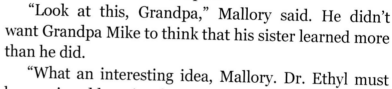

"Look at this, Grandpa," Mallory said. He didn't want Grandpa Mike to think that his sister learned more than he did.

"What an interesting idea, Mallory. Dr. Ethyl must have enjoyed learning from you and your sister as much as you enjoyed learning from her," Grandpa Mike said.

Melody and Mallory thought about that for a moment. Could they actually teach something to someone like Dr. Ethyl?

Mallory then told Grandpa Mike how Dr. Ethyl helped them to get an idea about big differences in size, by comparing big and little things they already know.

"Dr. Ethyl said that steam molecules are LOTS farther apart than water molecules. My drawing doesn't show how far apart they really are. Grandpa, do you know how much farther apart steam molecules are?"

Grandpa Mike shook his head "no."

Melody held up her thumb before Mallory could answer and said, "Steam molecules are 1,500 times farther apart than water. Dr. Ethyl showed us how to figure out how far apart that is."

Mallory found Dr. Ethyl's picture of a thumb and four cars. He was very happy to show Grandpa Mike (with some help from Melody) this interesting new way of figuring out how to get an idea of how big a difference in distance (or size) really is.

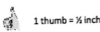 1 thumb = ½ inch

1,500 X ½ inch = 750 inches
750 inches/12 inches/foot = 62 ½ feet

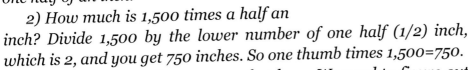

15 feet/car X 4 = 60 feet

1) We know how wide my thumb is: one half of an inch.

2) How much is 1,500 times a half an inch? Divide 1,500 by the lower number of one half (1/2) inch, which is 2, and you get 750 inches. So one thumb times 1,500=750.

3) Dad said lots of cars are 15 feet long. We need to figure out how many inches are in 15 feet.

4) Twelve inches are in each foot, right? So we have to multiply 15 feet times 12 inches. That's 180 inches for each car.

5) Next we have to figure out how many cars it takes to make up 750 inches, right? So we have to divide 180 inches into 750 inches and we get 4.166. That's four cars and maybe one front bumper.

"Whew!" That's some pretty fancy figuring," Grandpa Mike said. "Now you should be able to figure out all sorts of ways to compare big things with little things. Keep practicing and it will become easier each time."

Grandpa Mike seemed to understand that it took a lot of thinking to remember all these steps. Mallory told Grandpa Mike about George Mallory climbing Mount Everest. He looked up Mount Everest and found it is 29, 028 feet high. He divided 29, 028 by 15 and got 1,935.2 cars. He couldn't imagine that many cars.

"Try thinking about miles instead," his Dad suggested as he came into the living room. "One mile is 5,280 feet. Try dividing 29,028 by 5,280. How many miles is that?"

After doing some figuring in his notebook, Mallory said "it looks like it's about 5.497 miles, Dad. But how far is 5.497 miles?"

His Dad said, "Let's call it 5 and a half miles, more or less. That's about the distance from our home to the big shopping center."

"Wow! That's really HIGH!"

"It certainly is," Grandpa Mike and Mallory's Dad said together.

Some Geology

"Let me show you how geysers work, Grandpa," Mallory said. He turned to the page in his notebook that showed the four ingredients needed to make a geyser.

Mallory told Grandpa Mike that Dr. Ethyl said to think of his teapot as rocks in the earth that had cracks in it.

He explained how a geyser happens: you need water from above the ground (such as rain or snow), heat from underground, earthquakes, and the cracks in the rocks that form during earthquakes.

"It's kind of like a volcano, but with lots of water instead of fire and lava," Mallory said.

"And we saw different kinds of hot springs." Melody said. "Some have lots of water, and others have almost no water. They're almost dry. I don't remember what they're called, but they can change into a hot spring if it rains or water somehow leaks into them."

The 4 ingredients needed to make a geyser (think of Mal's tea kettle as underground rock with cracks)

Rain & snow

Vent (opening in the ground)

Harder rock underground

1. Water

3. Earthquakes cause 4. Cracks in rock

2. Magma (Molten rock)

Porous rock allows hot magma to rise through it

Mallory quickly looked it up in his notebook. "They're called fumaroles, Grandpa," he said, happy that he could find something that his sister didn't remember.

Some Biology

"Your Mom and Dad told me that you had a chance to look into a microscope," Grandpa Mike said after their Dad left the room.

Melody said, "It was really neat, Grandpa. Remember when we told you that water sometimes looks blue if the sky is really blue?

The paint pots were in lots and lots of different colors. Same thing on the ground. It looked like somebody wiped their paint brushes all over them."

"They were extremophiles," Mallory interrupted. "You can tell the temperature of the water according to its color."

"Is that so?" Grandpa Mike asked. "That's pretty confusing. Tell me what you learned about extremophiles."

Melody talked faster so her brother wouldn't have a chance to interrupt again.

"We saw some yellow ones in Dr. Ethyl's microscope. You can't see just one or two without a microscope, only when millions bunch together. Mom said extremophile means extreme loving. They live in extreme places, like where it's really hot or really cold. Pink ones like the hottest water, almost boiling. Greenish ones live next to them, in the next-hottest water. Then there were yellow and orange and other colors where the water got a little cooler."

"The extremophiles we saw like it really hot. Just like Mel. I think she's a big extremophile," said Mallory. He was anxious to tell Grandpa Mike his idea about his sister being a human extremophile, even though he wanted to be an extremophile too.

"Do you think you're an extremophile, Melody?" Grandpa Mike asked.

Melody shrugged her shoulders. "I don't know. I guess so."

Keeping in Contact

Grandpa Mike decided to change the subject. "Do your parents know Dr. Ethyl's full name?"

"She gave her card to Mom. I wrote down her name. Ethyl A. Shun, PhD," said Melody.

"I'm glad to hear that. You'll be able to thank her for all the time she spent with you and all the things she told you about. It's important to thank people for their kindness," said Grandpa Mike.

"We know, Grandpa."

Then the twins remembered one part of their Mom's "FEAST" advice for healthy living. "A stands for Attitude, and Gratitude is a very good attitude," she would say.

"Thank you, Grandpa! Yellowstone was really, really neat. We had a super time and learned lots of new things."

"You are both very welcome, children. I'm happy that this trip worked out so well for all of you, *and* for your parents too!"

"Oh, Grandpa. I forgot to tell you. I'm going to be an explorer, just like Mallory, the mountain climber!"

"Yes, Mallory. I remember a little bit about George Mallory. We'll have to find a good book on his life so we can both learn more about what he accomplished," Grandpa Mike said.

After Grandpa Mike left for home, Melody and Mallory asked their Mom if they could send an e-mail to Dr. Ethyl. "That's a very good idea," she said. "Let's do it together."

Next Adventure

When Grandpa Mike visited again a few days later, he said to Melody and Mallory, "Yellowstone was quite an adventure, especially for just a weekend. Your Mom and Dad think it would be a good idea for you to visit another new place, only closer to home this time. We've got a long weekend coming up at the end of May, Memorial Day. I'll be coming with you this time."

"Yay!" Melody and Mallory missed their Grandpa when they were at Yellowstone. They were happy that he would be coming with them this time.

"We'll be driving a special experimental car and you'll be meeting your Mom's cousin, Uncle Frank, for the first time. You'll also have the chance to meet lots of animals. Hm, is 'meet' the correct word to use for introducing people to animals?" Grandpa Mike asked.

"If we're driving, can we take Rufus?" Mallory asked, without even thinking about answering his Grandpa's question. He missed their dog almost as much as he missed Grandpa Mike. "I know our neighbors took good care of him while we were at Yellowstone, but I'm sure he was lonely when we were gone."

"Let's ask your Mom to phone Aunt Martha, just to be sure," Grandpa Mike suggested.

Grandpa Mike also hinted that on the way back home they might have a chance to get a close look at some old airplanes.

Melody's eyes got very big. Melody loves airplanes. "Old airplanes, Grandpa? Like the ones you told us about that you used to fly a long time ago?"

"Some are even older than that, believe it or not," he said as he gave her a wink and a really big smile.

As it turns out, Melody may not be the only extremophile in the family. The next book, *E – I – E – I – Uh-Oh*, tells how discoveries sometimes occur in ordinary places, in places that you might never suspect would reveal something extraordinary. The red star on the map on the next page shows where the Maloney family lives; The red circled numbers show:

(1) the location of Yellowstone National Park
(2) a family farm in southern Illinois, their next adventure.

48

INDEX

Acids, 35-36, 41

Bacteria, 7, 12, 27, 30-34, 37-38, 41
Bases, 41
Boardwalk, 8-18

Caldera, 17-18
Cauldron, 18
Celestine Pool, 14
Celsius, 25, 42
Comparing (big and little things),
    20-22, 42-43

Extremophile, 12, 27, 30-34, 37-39,
44-45

Fahrenheit, 26, 42
FEAST, 14, 45
Fumarole, 16, 27, 44

George Mallory, 38
Geothermal, 16-17
Geyser, 1-2, 4, 6-7, 10-11, 13, 16-17,
    19, 23, 25-26, 28, 43-44

H_2O, 20, 41
Hot spring, 6-7, 9, 12, 14, 16-17, 19,
    24, 27, 36, 40-41, 44
Hydrogen, 19-20, 33, 36
Hydrogen sulfide, 15, 19, 22, 41

Magma, 26
Microbe, 6, 30
Minerals, 12, 14, 17

Mud pots, 6, 9, 16, 24, 27

Octopus Spring, 27, 29-30, 34-35
Old Faithful, 1-7, 13, 20, 26
Oxygen, 19-20, 22, 33

Paint pots, 6-7, 9, 12, 16, 44
pH, 35-37, 41

Steam, 2-4, 6, 9, 15-16, 19-22, 25, 27-
    30, 32, 36-38, 40, 42, 43
Sulfur, 14, 22
Sulfur Cauldron, 36-37
Superheated water, 25, 27, 42

Tai chi, 7
Temperature, 25, 30-31, 34, 37, 44
Thermophile, 32

Volcano, 17-18, 44

Water, 1-3, 6, 9, 12, 14-17, 19, 21-22,
    25-30, 32-37, 41-45

Made in the USA
San Bernardino, CA
09 October 2015